MOHAMED EL BAKKARI

Tourism, local resources and human development

MOHAMED EL BAKKARI

Tourism, local resources and human development

The case of the sidi boulkhalf commune (Province of Azilal, Morocco)

ScienciaScripts

Imprint
Any brand names and product names mentioned in this book are subject to trademark, brand or patent protection and are trademarks or registered trademarks of their respective holders. The use of brand names, product names, common names, trade names, product descriptions etc. even without a particular marking in this work is in no way to be construed to mean that such names may be regarded as unrestricted in respect of trademark and brand protection legislation and could thus be used by anyone.

Cover image: Provided by the author

This book is a translation from the original published under ISBN 978-620-6-69992-7.

Publisher:
Sciencia Scripts
is a trademark of
Dodo Books Indian Ocean Ltd. and OmniScriptum S.R.L publishing group

120 High Road, East Finchley, London, N2 9ED, United Kingdom
Str. Armeneasca 28/1, office 1, Chisinau MD-2012, Republic of Moldova, Europe
Printed at: see last page
ISBN: 978-620-8-28688-0

Contents

General introduction

Morocco's mountains are rich in biodiversity, but unfortunately most of their inhabitants live in poverty. This is because crucial natural resources, such as forests, minerals and water, are frequently appropriated and exploited by foreign actors. Mountain dwellers rarely benefit from these resources, and economic activities in these regions remain largely underdeveloped, with little sustainable or rational management. This situation leaves mountain communities facing persistent financial difficulties.

On the other hand, mountain areas have been neglected in the past, receiving neither the attention nor the public investment needed for their development. At present, these mountain areas are lagging significantly behind in terms of socio-economic and cultural infrastructure, and the living conditions of many local populations remain unsatisfactory overall.

Unfortunately, this pessimistic view is widespread in most developing countries. However, Morocco has major assets that could play an active role in local socio-economic development.

This underlines the fact that Morocco's mountainous regions are not only repositories of important natural resources and valuable local know-how, but also represent real opportunities for development. These areas are the home and place of activity of the indigenous population, playing a crucial role in the development of these territories.

Despite all their assets, the situation in Morocco's mountain areas remains very difficult today. These areas suffer from major climatic and geographical handicaps, which have been exacerbated by their marginalisation. Although several initiatives have been launched in these areas, they have not had the desired effect. They have not had the necessary scope to contribute to a real integration of this part of the territory into the national economy. Often, the advantage of a logic of short-term financial profitability has led to the acceptance of regional disparities and increased poverty. It is the opposite of this logic that will enable these areas to be re-established in their economic, cultural and historical role and in their role as a territorial and ecological barrier.

With this in mind, the upgrading of mountain areas and human development are becoming imperatives for preserving these ecosystems, which are of vital importance to the national economy.

While the National Initiative for Human Development launched by His Majesty on 18 May 2005 provides an explicit political and institutional framework conducive to the implementation of development actions, it calls on us all to put it into practice and strengthen it.

The experiences available in Morocco were all based on the idea that it is the State which, on the basis of a pre-conceived model, should define the guidelines and decide on the most appropriate actions and the way in which these actions should be carried out. Such a centralised, top-down approach does not allow rural populations, and more particularly the poorest groups, to participate in the choice of development programmes and their implementation.

more than 9a, the State has positioned itself as being in a position to define the needs of the population and to decide on the actions required to satisfy them, thereby subjecting these populations to a form of subjection and impotence.

Until the early 90s, the planning and management approaches recommended did not achieve the sustainability of the natural resources themselves, particularly in marginal agro-ecological zones, or of the improved farming methods proposed. As a result, these areas are increasingly threatened by "human desertification" and land abandonment (SDR 2020).

Awareness of the scale of the failures is increasingly driving development efforts towards participatory, bottom-up approaches, which are proving to be more effective than the top-down approaches used in the past.

More than ever, integration and participation are the key principles of any community development project or action planning process. Although efforts have been made by various projects throughout the Kingdom, there are still differences of meaning and implementation. We need to stop and take a look at what has been achieved so far, in order to move towards a realistic, flexible approach that can be adapted to the diversity of the areas in which we operate.

The participation of populations in development actions has become an imperative demanded as much by government structures working in the field of development as by donors. In terms of regional development policy, participation is a guiding principle and a sine qua non for the implementation of national guidelines.

In the field of rural development, "participation" has become part of the official language of most stakeholders. Its application involves a change in the perception of the role of rural populations and in the methods of action.

The Moroccan public authorities are counting on the active and effective participation of rural populations in order to initiate and maintain local development. With this in mind, a large number of development initiatives are being implemented throughout the Kingdom, but the 1st observation is the low level of support from rural people for these initiatives, due to a number of constraints currently affecting Morocco in terms of participation in rural development and the protection of natural resources.

It is with this in mind that the present work sketches out a study of local development in mountain areas - constraints and potential - the case of the Sidi boulkhalf commune.

The final objective of this study is to formulate a development project tailored to the specific characteristics of this rural community, using a participatory approach.

This work is structured in three main parts:

The first chapter is devoted to the problem and the research methodology.

The second chapter presents the potential and constraints of the sidi boulkhalf commune.

The final chapter (chapter 3) aims to propose a territorial project that reflects the specific characteristics of the Sidi Boulkhalf commune.

In the light of the lessons learnt, the reflections and analyses above, a general conclusion and recommendations will sum up our research.

I Chapter

Problem and research methodology

Introduction

The situation of mountain areas in Morocco remains very difficult today. It is marked by phases of marginalisation and economic dëvitalisation that continue to accelerate. These areas suffer from major climatic and geographical constraints, which have been made worse by their lack of access. Although several initiatives have been launched in these areas, they have not had the necessary scope to contribute to a real integration of this part of the territory into the national economy.

Despite all the riches offered by natural resources and cultural heritage, the sources of income for people living in mountain areas are still limited to agriculture, livestock farming and forestry products.

Furthermore, the introduction of a local development process emanating from "below" presupposes the existence of decision-making capacity and responsibility at local level, something that remains largely dependent on a real decentralisation of the political-administrative system and the application of the principle of subsidiarity. What is the reality?

1. Problem :

In Morocco, as in all developing countries, development is a major concern, and rural development is particularly important, firstly because almost half the population lives in rural areas, and secondly because of the disparity between urban and rural areas.

Today's rural world is faced with a number of problems that are real obstacles to its development. Social characteristics, the diversification of the rural population and strong demographic pressure are decelerating the development process in rural areas.

Agriculture, the main source of income in rural areas, is not yet able to realise its potential in terms of productivity and job creation. The inadequacy and precariousness of economic and social infrastructure in rural areas is also reducing the capacity for agricultural reductions and the initiatives of the rural population.

The only consequence of this was that the rural world was strongly marked by the poverty and vulnerability of its poor population.

However, the population's participation in the development process is severely limited by the lack of professional organisations and associations, which inhibits the overall dynamism of rural society.

The problems facing the rural world call for a national strategic approach to find workable solutions, with the participation of all the players involved in this development.

The rural commune of Sidi Boulkhalf (the subject of our study) also belongs to the group of communes suffering from the problems and barriers to development mentioned above.

After an initial look at the history of the town's development, it became clear that there are major challenges to be faced, structural problems that need to be remedied and strategic issues that need to be addressed.

It is against this backdrop that our study will attempt to help answer the following question:

What local development strategy is appropriate for the rural commune of Sidi Boulkhalf?

To understand this central issue, we believe it is essential to answer the following questions:

1- What constraints are hampering the development of Sidi Boulkhalf?

2- What resources does the local authority in question benefit from?

3- What projects should be developed to meet the specific needs of the Sidi Boulkhalf commune as part of a participatory approach?

2. Research methodology :

Scientific disciplines always start with hypotheses, develop methods, carry out experiments and produce results. There is no doubt that obtaining accurate and logical results is closely linked to the working method adopted.

According to PASCON (1981), "there is no single, all-purpose method; every search is first and foremost a search for a method".

Thus the choice of dëthodes depends mainly on the objectives of the study and the resources available.

3. Information gathering tools :

3- 1 Interviews

In order to gather as much information as possible about our study, we have chosen a semi-directive interview as our investigative tool. We believe that this is the most appropriate means of providing answers to the questions we have posed in the context of our problematics.

GRAWITZ (1990) reports that "it is a procëdë of scientific investigation, using a process of communication, to gather information in relation to the goal ňxë"

This investigative tool also allows us to avoid the risk of leaving unanswered ëlëments that could be of great importance to our study.

In our study, the interview allowed us to learn about the role of the rural commune in dëvelopment by trying to analyse with the interviewës the actions that have ëtë effective by the rural commune Sidi boulkhalf. The interview also allowed us to grasp the different factors that can influence the participation of stakeholders in dëveloppement (local kins, local population, commune staff, investor, and local autor^s).

Unlike a questionnaire survey, an interview is characterised by direct contact between the researcher and the interviewee and by little direction from the researcher. Thus the idea of getting the interviewee to talk, of putting him in a position to express himself at greater length and more freely, of spontaneously saying things that come to his mind by association of apropos ideas, sometimes unrelated to the theme ttborek.

We chose the semi-directive interview as the type of interview. It is directive in the sense that it is neither entirely open nor channelled by a large number of precise questions.

3-2 Participatory observation :

The observer, integrates to varying degrees into the situation, in order to understand it better" (COPIN, 1998 e'Пë by ABELOUCH, 2006)

Participatory observation is a technique gënëralement utilisëe par les sociologues et les anthropologues, implique la présence et la participation de l'observateur dans la vie sociale et culturelle de population ëtudiëe. It is visual or auditory.

3-3 Analysis method :

Analysis is a continuous process of reviewing the information collected, vërifying it and drawing conclusions. In our study, the information thus obtained was subjected to a process that includes:

- **A descriptive analysis**: this analysis enables us to present and describe the situation in a clear and simplified way, and to gather the information we need to carry out our investigations.
- **Continuous analysis**: this analysis enables the information gathered to be filtered and selected, as well as an objective search for the clear continuity of the communication and the logic of the players involved.
- **A comparative analysis**: the aim of this analysis is to compare the results of the investigations obtained with other similar experiences and to draw certain conclusions.

4. Research hypotheses :

The main hypothesis guiding this research is that local development in the rural commune of Sidi Boulkhalf requires the participation of all the elements that make it up, namely civil society, local authorities, local elected representatives and the private sector.

This main hypothesis gives rise to the following specific hypotheses:

- The municipality of Sidi Boulkhalf needs to make the most of its resources and potential if its development is to take off.
- The rural population participates in local development projects and programmes.

5. The aim of the study :

The aim of this study is to analyse the main potential and constraints of the Sidi Boulkhalf commune, based on the projects already in place, and to make a number of recommendations to identify the preconditions for designing a suitable territorial project based on a participatory approach that meets the specific characteristics and needs of all the elements of the commune in question.

This theme has been chosen for the following reasons:

The scarcity of studies on rural communities.

Awareness of the local development issue among local communities and civil society has increased in recent years.

(local governance, transparency, territorial approach, etc.)

II chapter

The Sidi boulkhalf, enormous potential but also but also constraints to local development

Introduction

Since the 1990s, Morocco has been overhauling its rural development strategy. In the face of accumulated social, economic and political backwardness, this strategy seeks to refocus state intervention in favour of development that reflects a comprehensive, integrated and participatory approach.

The aim is to create a favourable environment for correcting imbalances and developing the potential of rural areas. The key lever here is the creation of clusters of projects tailored to the realities of the target areas, with a view to promoting the emergence of development poles.

So, in this chapter we are going to present the different components of the Sidi Boulkhalf commune (the subject of our study), with the aim of identifying the potential and constraints of the commune in question, in order to know the general framework of local development which corresponds to the requirements of the environment and the expectations of the population.

I - General presentation of the commune of Sidi Boulkhalf

The commune of Sidi Boulkhalf was created on 16 October 1992. It is located in a mountainous area and its population is drawn from the Oueltana tribes. Its geographical limits are as follows:

To the south by the two communes of Ait Tamlil and Ait Boualli,

To the east by the two communes of Ait Blal and Ouaoula,

To the north by the municipality of Ait Majden,

To the west by the municipality of Tifni.

The municipality has 4 Mashiakhats:

Mashiakhat Iouaridene,

Machiakhat Ait Ziad,

Machiakhat Ait Toutline,

Machiakhat Amalou.

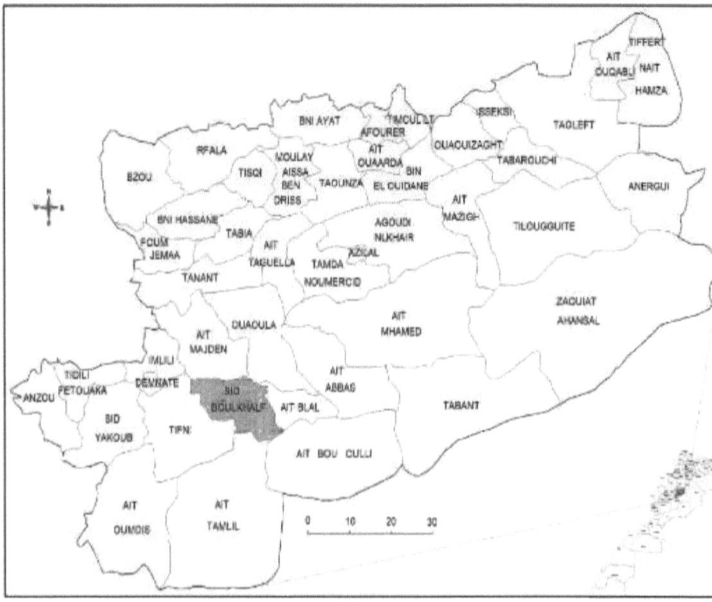

Figure 1: The дёпёга1е situation of the sidi boulkhalf commune.

Located in the south-west of the Province of Azilal, the Sidi Boulkhalf commune covers an area of 167km^2 and is characterised by its uneven, enclosed topography; it is cut by the valley of the Oued Ghazzef, as well as other small wadis such as Arga N'Tarit, Ait Toutline, Immi n'ouaka and Zrite, and around ten water springs. The commune is crossed by a section of the RR 302.

II - Potential and constraints of the municipality

1 - Human potential, a major asset for the commune exposed to external influences

Knowledge of the characteristics of a population is of great importance for studies aimed at improving the population's standard of living in order to determine its various needs in different areas (food, schooling, employment, health, housing, etc.). For example, the age structure provides information on the demographic make-up of a population and on the evolution of its trends and their significance. It also provides valuable information for planning the needs of the population in different socio-economic sectors.

13149 inhabitants in 2004

This is a medium-sized commune in the Demnate district, and a large one in the province, accounting for 2.6% of the total population.

Table 1: General demographic situation

RGPH 1982	RGPH 1994	RGPH 2004	Estimation 2013
8268	11174	13149	15964

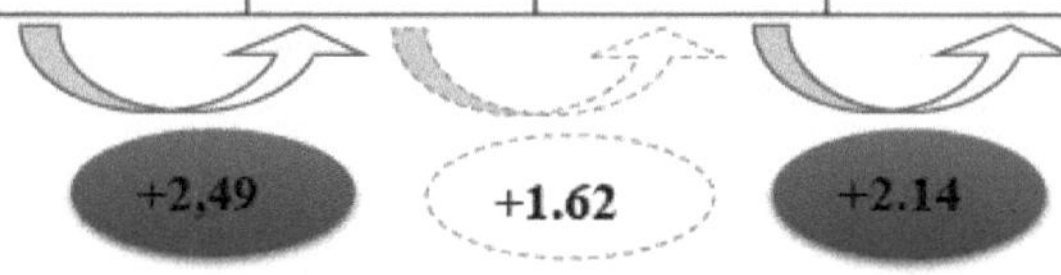

The municipality of Sidi Boulkhalf saw its average annual net population growth fall between the periods 1982-1994 and 1994-2004, from +2.49% to +1.62% and +2.14% in 2013. This situation can be explained firstly by the rural exodus phenomenon and secondly by family planning efforts. In fact, the natural growth rate of 3.99% for the period 1994/2004 is much higher than the net growth rate for the same period, which is only 1.62%, which could mean that the number of people leaving the country is higher than the number of new arrivals.

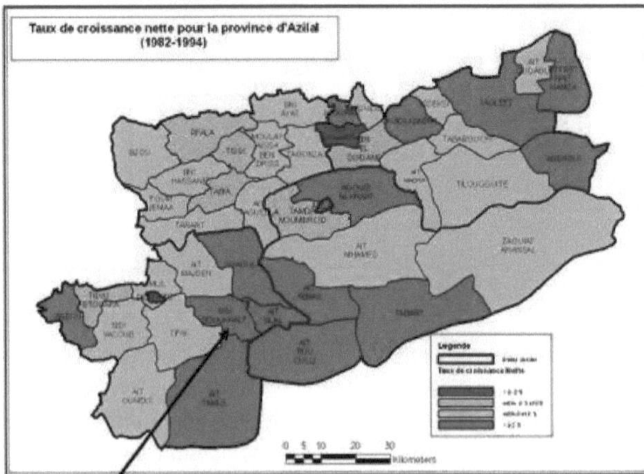

Figure 2 : taux de croissance nette pour la province d'Azilal (1982-1994) sur la base des données de RGPH 2004

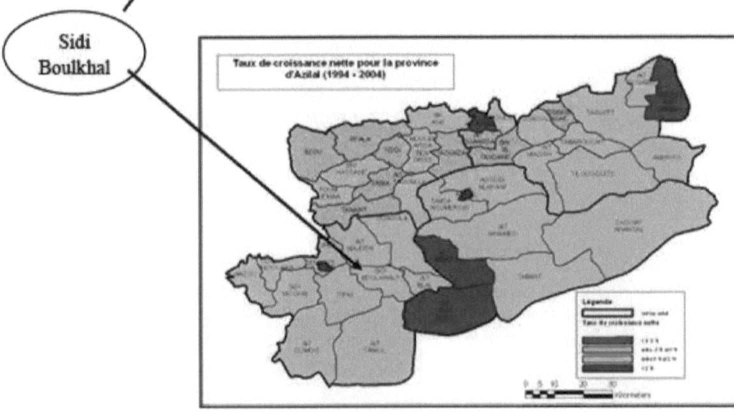

Figure 2: *Net growth rate for the province of Azilal (1982-1994) based on RGPH 2004 data*

Figure 3: Net growth rate for the province of Azilal (1982-1994) based on RGPH 2004 data

1 - 1 Population distribution in the municipality

The population density in this commune was around 78.4 inhabitants/km^2 in 2004, higher than the national average of 45 inhabitants/km2 and the provincial average of 50.2 inhabitants/km2.

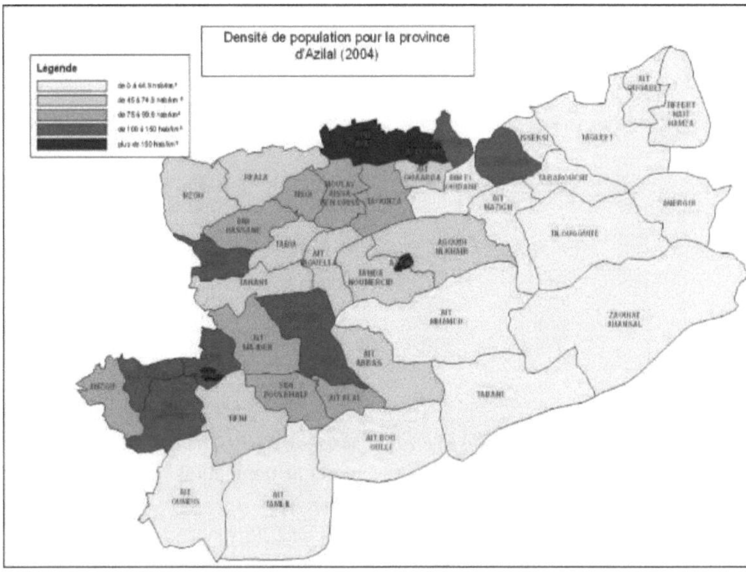

Figure 4: Population density map for the province of Azilal, based on data from

RGPH 2004

The General Population and Housing Census carried out in 2004 identified 44 douars in the sidi boulkhalf rural commune, with 11% of the population concentrated in the two douars Tangakhte (close to the Ait Bllal- Imminifri road) and Taghbaloute (the commune's administrative centre, where most of the basic infrastructure is located). The douars with the highest negative growth rates were: Ait Sigmine (-3.43%), Tagnite (-7.47%), Imzizoine (-7.17%), Ghobdade (-3.27%), Amalou (-2.34%) and Ilaguate (-2.14%) (douars located in the mountainous area, where there is a lack of basic infrastructure).

In addition, other douars are experiencing a positive increase in their population growth rate, such as: Oukta +7.16%, Ikhachene +6.53% (both douars are located in a remote mountainous area, hence the lack of family planning, and management at a young age), Tilsakhte +2.95% (douar close to the road).

1 - 2 Population structure :

1-2-1-Distribution by age group and sex in 2004 :

Table 2: ***Breakdown of data by age group and sex of RGPH 2004***

Classes	Men	Women	Total	Share of women	Share of each class
0-5	1123	1066	2189	48,6%	16,6%
6-14	1815	1612	3427	47%	26,6%
15-59	3244	3466	6710	51,6%	51%
60 and over	465	358	823	43,4%	0,6%
Total	6647	6502	13149	49,4%	23,7%

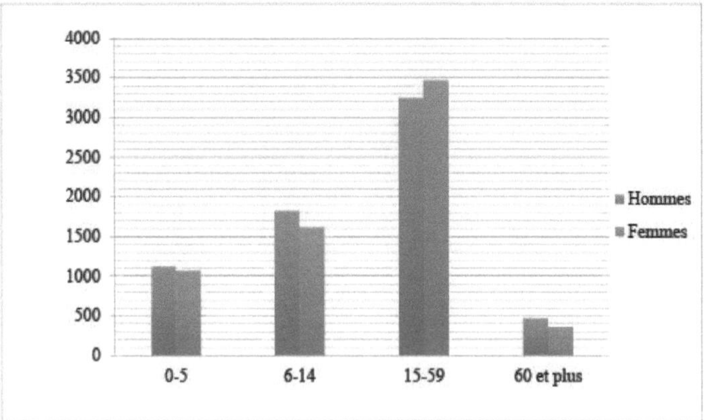

Figure 5: Age and sex groups, based on RGPH 2004 data

A simple reading of the table shows that the working population represents 51% of the total population. 26.6% of the population is of school age.

<u>1-2-2Average annual net growth rate between 1994 and 2004 :</u>

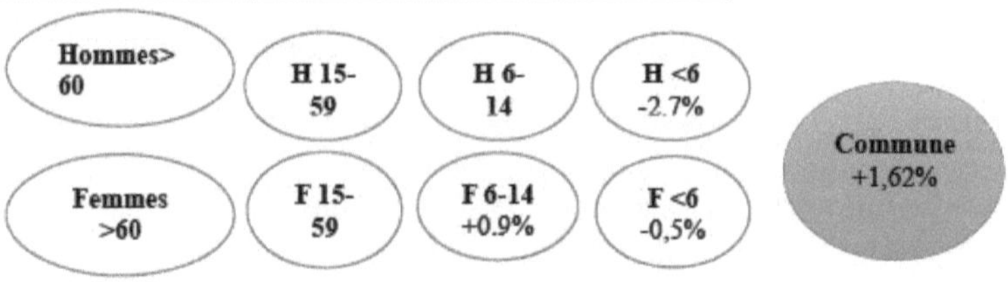

This breakdown can be derived from the following assumptions:

Table 3: Assumptions on the demographic distribution of the population

M and F >60	No emigration, personal stability
H 15-59	Seasonal emigration to cities for work (construction, informal sector, etc.)
F 15-59	Personal stability
M and F 6-14	- Dëmographic transition (as the under-6 age group shrinks, this is gradually reflected in the next group of under-14s) - Impact of family emigration, which affects children more than parents, because family emigration affects both parents on the one hand, and 3 or more of the children on the other. - Late management
M and F <6	-Demographic transition (family planning chosen by families to reduce expenditure on children in a context of falling agricultural incomes) - Late management

1-3 Migration flows

<u>Table 4</u>: Migration flows in the municipality, based on ARP Developpement 2009 data

MIGRATORY	EMIGRATION

FLOWS	INTERNAL		FOREIGN	
	INDIVIDUALS	FAMILIES	**INDIVIDUALS**	**FAMILIES**
NUMBER	**2543** Mostly men	**472**	**29**	**45**
WHO IS IT?	Young people between 20-35 (90% of them have a primary school education)	Most families left their douars in the 80s.	Young people between 20-35	Most of the emigrants left their douars in 1973.
WHERE?	- Agadir : 984 - Casablanca: 643 - Marrakech: 576 - Flap : 3 - Other : 139	- Agadir: 120 - Other: 174 - Demnate: 83 - Marrakech: 20 - Casablanca: 17 - Flap : 3	- France :15 - Spain: 7 - Italy: 7	- France: 40 - Spain: 5
	92,5%		7,5%	
FACTORS	Decrease in income generated by agricultural activity (particularly due to droughts).		- Lack of local job vacancies - Insufficient basic social services	
	Inactive during the dry season. Isolation of certain douars (Ikhachene, Oukta... Lack of local job offers. Young people can no longer work in the agricultural sector.		- Weak income generated by local activities	
WHY DO IT?	**Looking for work:** - BTP : 1405 -Agriculture : 918 - Trade : 10 - Other : 40		- **Looking for work** - **Lifestyle improvements** - **Looking for work** in 1 agriculture, construction and trade	
WHEN?	Emigration of j eunes throughout Γannëe, returning every two or three months and during religious occasions, sending a monthly amount varying between Dhs 250-300 to family members.		Individual emigrants have left the douars in droves	Immigrants from Γ dtran ger have left the douars for good and only return during the summer holidays, most of whom live in Demnate.

According to this table, around 78% of the active male population work outside the country, and around 80% of emigrants have a primary education and no vocational training.

Most of this emigration is within the country (92.5%), with Agadir and Casablanca the most targeted cities.

2- The Sidi Boulkhalf commune has enormous natural potential for local development:

2 - 1 Water resources :

From a hydrological point of view, the study area consists of a basin with significant hydraulic potential.

Most of the water resources in this commune are found in the valley of Oued Ghazzef, as well as other small wadis such as Arga N'Tarit, Ait Toutline, Immi n'ouaka and Zrite, and a dozen or so springs. These resources are used by the local population to meet their water needs, for domestic use, for watering livestock and for irrigating farmland.

Photo 1: *View towards the north-east. Irrigated land located 3 km east of douar Taghbaloute - the centre of the Sidi boulkhalf commune. Irrigated agriculture has developed on the land at the bottom of the Ghazef valley. This land is used primarily for growing fruit trees (vines, apples and walnuts), followed by fodder crops, mainly alfalfa.*

2 - 2 A dominant agricultural activity subject to constraints

2 - 2 The agricultural sector :

2 - 2 - 1 Environmental conditions :

Sidi Boulkhalf is an agropastoral commune, with a population that is very attached to its land. Despite the many difficulties associated with farming and livestock rearing - the mountainous terrain is unfavourable for agriculture, and poor "Hrach" (70%), "Byad" (20%) and "Hamri" (10%) soils predominate - these are the most important economic activity, enabling a fringe of the population to continue living in the area.

Rainfall in the commune is irregular, varying between 450 and 500 mm depending on altitude. There is snow in the douars at high altitude (over 1500 m) for a period of 4 months (December-March).

Water resources in the Sidi Boulkhalf commune generally come from oued Ghazzef and springs, which dry up during the rainy season.

2 - 2 - 2 Land and its distribution among farms :

The land in the Sidi Boulkhalf commune has a total surface area of 54,651 hectares ([1]),

[1] [1] Monograph on the municipality

including 5,161 hectares ([2] , 9.4%) of UAA, 11,273 hectares (20.6%) of woodland and 550 hectares (1%) of rangeland and uncultivated land. Bour crops account for 87.19% of the UAA, while irrigated land accounts for 767 hectares (14.8%) of the UAA, including 623 hectares under permanent irrigation and 144 hectares under seasonal irrigation.
There are only Melk plots, which account for 83.03% of the UAA, state-owned plots (16.84%) and a few Habous plots (0.08%). The most dominant form of tenure is direct (98.4%), followed by lease in kind (1.14%) and rental (0.42%).

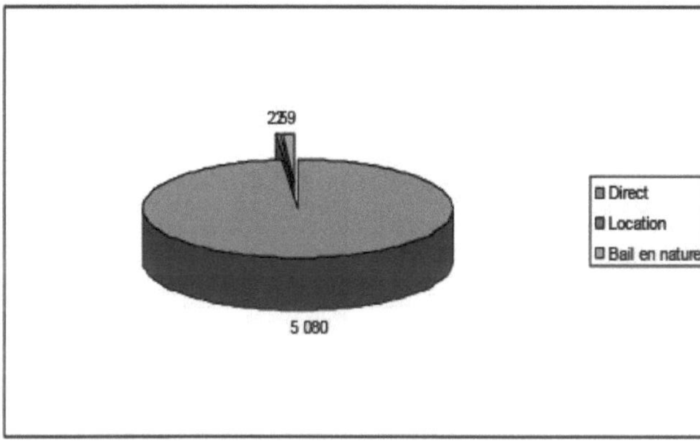

Figure 6: Land tenure based on data from the municipality's monograph

In 1996, the average size of UAA per farm was around 3.9 ha, compared with 4.1 ha at provincial level, and the average size of plots was 0.3 ha. The farms are characterised by the fragmentation of plots of land, the fact that they are in short supply due to inheritance tax and the fact that they are not registered. The number of plots per farm varies between 10 and 20.

2 - 2 - 3 Other factors of production :

There is a permanent family workforce, including children of school age (10 to 16). There is a shortage of agricultural labour due to the emigration of young people who do not want to work in agriculture.
Given the commune's mountainous terrain and the fragmentation of plots, most agricultural work is not mechanised, with the plough used for tilling the soil, while harvesting is always done with a sickle or by hand, except in certain cases where the ears are threshed using threshing machines rented from the Attawiya souk at 120 Dhs/hour.
According to the 1996 RGA, the use of fertilisers was high (87%), but farmers stated that this use is becoming low because of the increase in their price and the succession of drought years, although they do use manure from livestock activities. In addition, there is a total lack of awareness of the use of selected seeds and plant protection products (0% and 2% of farms respectively in 1996, compared with 5% and 11% at provincial level). Most farmers obtain their supplies from the Demnate souk or from previous harvests.

[2] RGA 1996

Photo 2: The seguias used for irrigation are very vulnerable and need to be improved and maintained due to climatic and topographical constraints.

Irrigation is mainly provided by seguias from wadis and springs, most of which are not amënagëes.

<u>*2 - 2 - 4 Production techniques :*</u>

- <u>I . Technical itinerary of the cërëales :</u> Shallow ploughing begins in October in mountainous areas at medium altitude, and in January in mountainous areas at high altitude (snowfall), it is generally done with an animal-drawn plough and by sowing for sloping plots that have not been cleared; only animal manure (1.5 q/ha) is applied as a base fertiliser.

Sowing is done by hand at a rate of 1 quintal/ha, the seeds are brought in from the weekly souk at a rate of 3 dhs/kg, so the cost of sowing for one hectare is 300 dhs; and manual weeding is done during the March-April period, in order to use these grasses as cattle feed. As for harvesting, this is done manually from April onwards in mountainous areas at medium altitude, and in June in mountainous areas at high altitude.

- <u>Technical itinerary for fruit trees:</u> The fruit trees are mainly olive, almond, carob, apple, vine and walnut:

- **For the olive tree:** Farmers plant 15-20 cm seedlings simultaneously with cereal, fodder and vegetable crops; these seedlings are bought from the Demnate souk. The use of chemical fertilisers is reduced to a minimum; traditional manure-based fertilisers are widely used. The olive trees are attacked by a number of pests such as psyllids, although no phytosanitary treatment is applied, and the harvest begins in December.

- **The almond and carob trees are** sown with seeds buried 2-3 cm deep from previous harvests, but no fertiliser is applied and no plant protection treatments are applied to these two crops because of the high cost of plant protection products and the lack of technical support for farmers. No maintenance is carried out on these trees, and the manual harvest begins in December.

- **For the walnut tree:** Sow the seeds 4-5 cm deep in September; transplant the trees at the age of 3 years and 1.5 m in length into rootballs; apply 1.5 kg of animal manure per tree in the first year of planting; the trees do not have to be pruned. The harvest takes place in July.
- **Concerning the apple tree:** It is not pruned, in the knowledge that phytosanitary treatments are not applied, and moreover the harvest takes place in September-October.
- **For the vine:** Planting is done from cuttings (10-15 cm long) taken from previous plantings in March. Phytosanitary treatments and fertilisers do not apply and the harvest is done from September.

2 - 2 - 5 Crop rotation and technical performance

Cereal crops dominate (4,475 hectares), making up 86% of the UAA. As far as cereals are concerned, barley (used as animal feed) dominates, followed by durum wheat and soft wheat. Cereal production fluctuates from year to year depending on rainfall. This crop is directly affected by climatic hazards. For example, the average yield for the 2006-2007 season was 15 quintals, but fell to 3 quintals the following season.

Arboriculture is dominated by walnut and almond trees. Carobs, apples, vines and olives are also grown in mountainous areas, and cereals are always grown alongside fruit trees.

Olive cultivation is based on the "Moroccan picholine3 " variety,[3] , which is expanding rapidly in the medium-altitude mountain douars, due to its low production costs and its adaptation to the conditions of the area, especially its hardiness and low maintenance requirements. The area given over to growing carob, walnut and almond trees has increased in recent years, a trend that could be explained by the income they provide to farmers. But the problem of annual frost for a period of 4 months (December-March) is causing serious damage to crop yields. Not to mention the flooding of Oued Ghazef, which causes damage to neighbouring crops.

Fruit tree production varies according to the age of the tree, and is summarised in the table below

Table 5: **Fruit tree yields, based on data from an interview with a technician from the Demnate CTA**

Speculation	Years of production	performance
Almond tree	3rd 15th 20eme	0.25 Kgs/tree 6 to 8 Kgs/tree 1 Kg/tree
Olivier	^eme 10th	0.25- 0.5 Kg/tree 20- 30 Kgs/tree
	120th	1 Kg/tree
Carob tree	5th year	40 to 50 Kgs/tree
Vine	4eme year	10-15 Kgs/tree
Apple tree	1st year	0.5 Kg/tree

[3] Interview with a technical officer at the Demnate air traffic control centre

	4th year 6th year 10th year	4-5 Kgs/tree 20 Kgs/tree 150 Kgs/tree
Walnut	Year 12 30th year	3-4 Kgs/tree 80 Kgs/tree

The sector irrigates:

The traditional irrigation sector (small and medium-scale hydropower) covers an area of 767 hectares, of which 623 hectares are under perennial water and 144 hectares under seasonal water. This sector is made up of three areas:

> Tamguert: 502 Ha
> Ait Toutline: 85 Ha
> Immi Nouaka: 120 Ha.
> Ikhachane: 60ha.

Within these PMH perimeters, fruit growing comes first with 460.20 Ha, followed by forage crops (230.10 Ha), mainly alfalfa and to a lesser extent maize, cereals: barley and durum wheat (115.05 Ha), market gardening (28.76 Ha). And leguminous crops: fodder and peas (23.01 Ha).

It should also be noted that all agricultural produce is used mainly for own consumption, except for vines, carob trees, apple trees, almond trees and walnut trees, which are sold.

Walnut trees are sold to traders from Marrakech at 50 dhs/Kg and some farmers sell them dëcortiquë at the Demnate souk at 70-100 dhs/Kg. Almond trees are sold dëcortiquë at between 40 and 150 dhs/Kg.

Apple trees are sold to traders from Merrakech at an average price of Dhs 5-10 per kg or at the Demnate souk at a price of Dhs 7-15 per kg.

The Caroubier is sold at the Demnate souk at a price of between 5 and 10 dhs/Kg. The vine is also sold on the tree to traders from Marrakech at a price of between 6-10 dhs/Kg.

When production is high, the olives are sold at 6-10 dhs/Kg in the Demnate souk. In most cases, the olives are crushed to extract the oil in the traditional maasras, and payment is made in two ways: either at 0.70 dhs/Kg of olives, or by deducting one tenth of the quantity of oil obtained. Olive oil is mainly intended for self-consumption. There is a lack of suitable storage facilities for the packaging of agricultural produce during periods of high vine and apple production.

Table 6: Number of livestock in the commune of Sidi Boulkhalf, based on data from the Azilal DPA veterinary service, 1996.

Year	Cattle	Sheep	Goats
1996	2.181	8398	11478
2013 estimate	1600	6.000	5000
Growth rate	-40%	-36%	-60%

A simple reading of this table shows a very sharp fall in numbers between 1996 and 2013, of 36% for sheep, 40% for cattle and 60% for goats. This may be due to the sale of livestock because of the lack of grazing areas and the increase in feed prices.

The breed of cattle is crossbred (local*Holstein); it is mainly used for milk production for self-consumption. The breed of sheep is crossbred (local*Sardi); goats are a local breed,

2 - 2 - 6 Breeding technical systems

The cattle are always kept in stalls, especially in the immediate vicinity, and the goats are grazed in the forest on the herbaceous layer throughout the year, except during the snowfall for a period of 4 months (December-March), when tree branches and shrubs are used as feed. Cattle and sheep are fed during the October-April period, mainly bran, barley, straw, alfalfa and maize fodder. Sheep are also grazed on fallow land, rangeland or on the edges of plots during the May-August period.

2 - 2 - 7 Water and grazing land management

The irregularity of the rainfall has a major influence on the areas available for grazing and the natural watering places, and the herds are watered either from the seguias or from spring water.

In the event of livestock illness, farmers are responsible for the costs of treating their livestock. They have to travel as far as Demnate, and visit the vëtërinaries during the vaccination campaigns, which take place 1-2 times a year in April and September.

2 - 2 - 8 Upstream/downstream analysis of production sectors

The obstacles to the development of livestock farming are mainly related to the feeding of goats, the hunger gap period (October to February) is difficult, given their dependence on very poor forest rangelands during this period, the reduction in pastoral resources from rangelands and forests following the succession of drought years and over-ripening, as well as the absence of commercial outlets for the sale of milk due to the difficulty of access to douars and their remoteness, the lack of structure in the sector (few professional organisations), the lack of manpower to look after the herds, especially goats, which is done in the forest rangelands, family manpower for this activity is becoming increasingly scarce (ëcol e or ëmigration of young people to the towns) and the lack of grazing land in the mountain douars.

2 - 3 Degraded forest resources

The wealth of water resources and the existence of rugged terrain have given rise to a varied vëgëtation with significant forest cover,

The forest area which occupies a large area, of which the forest represents 11 273 Ha or 20.6%, the agricultural area represents 5161 Ha or 9.4%, however it knows a massive exploitation from the neighboring population which leads to the dëgradation pei'iretueile forest heritage whose racesste to consider interventions to safeguard the wealth йм^йёге.

Such an area plays a number of economic, social and ecological roles. Many local people use it to raise goats.

Sidi Boulkhalf is a commune with very significant forestry potential (11273ha). Nevertheless, this potential is largely overexploited: wood harvesting and overpopulation leading to the degradation of the resource and the risk of erosion. The management of these vital areas for the municipality needs to be restructured and rethought between the various stakeholders.

This first guideline therefore calls for concerted reflection on how to better manage silvo-pastoral resources. Active measures could be defined both in terms of protection and reforestation, the definition of a management framework and the promotion of alternative solutions to firewood or livestock feed.

Photo 3: The overexploitation of wood resources to meet demand for traditional energy poses a major threat to forest resources, which are already growing at a low annual rate (estimated at 0.16 m^3 /ha/year). As a result, the vegetation cover of many areas is transformed each year into wood and charcoal, leaving bare land exposed to all forms of erosion.

***2 - 4* Iouariden tourist site: an exceptional heritage to be protected and developed**

In terms of tourism, the Sidi Boukhlaef commune has enormous potential for becoming a tourist attraction. It is characterised by a contrasting and varied natural environment, enhanced by the existence of a natural heritage that is one of the prides of its inhabitants and a specific asset. The mountains and plains bordering the commune are two landscapes that have a major influence on tourism.

Dinosaur footprints

The Iouariden basin is famous for the numerous dinosaur footprints imprinted in the red detrital sediments known as the "Gres Guettioua" or red Dogger layers. The dinosaur footprints are located in red clays with desiccation figures, indicating a lagoon-like or sebkhaic continental environment with temporary emersion.

The iouariden region includes three major dinosaur footprint sites:

- Iroutlane-Iwariden
- Ait Mimoun Ufellah
- Jebel Bruna

There are three kinds of tracks represented by several shapes. The large, oval tracks are those of a large, heavy herbivorous dinosaur with the scientific name of Breviparopus taghbaloutensis, the genus is Breviparopus, the species is Taghbaloutensis; the latter name comes from the small hamlet where these tracks are found, called taghbaloute, which means spring in Tamazight. The other dinosaur is a tridactyl (three-fingered), its tracks resemble bird feet, and it is a carnivore, (Jacques Gandini).

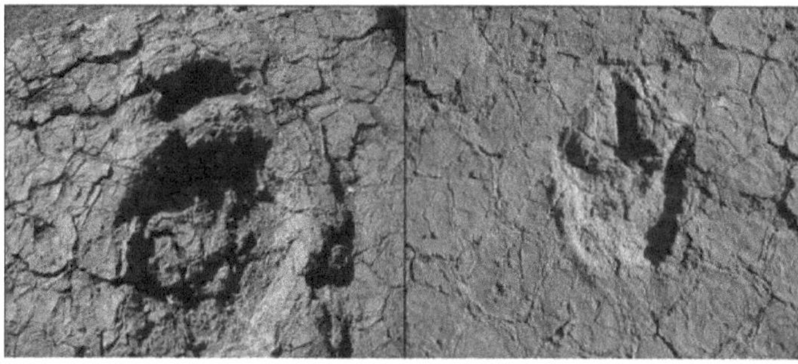

Photo 4: Dinosaur footprints imprinted in red detrital sëdiments offer enormous potential for tourism.

<u>State of conservation:</u>

The dinosaur footprints at Iouaridenes are in a poor state of conservation and are changing due to their exposure to a multitude of damaging factors:

J atmosphëric factors: rain, insolation, wind, hydrolysis, frost and thermal oscillations (water seeps into the cracks and pores of the rock. As it freezes, the volume increases, leading to the fragmentation of the rock);

J Biological factors: roots grow and increase in the cracks, leading to the dësagrëgation of the rocks. Visits by tourists arouse the interest of the local population, who use these deposits as meeting places, leading to dëtërioration.

3- The commune of Sidi Boulkhalf is subject to a number of constraints that must be taken into account in the commune's development project

3-1 Human resources threatened by immigration

The demographic analysis has shown us that despite the existence of a young and active population which constitutes a reserve of life force to the extent of 51% of the total population, the commune is not yet sheltered due to several constraints since 78% of the active male population is temporarily migrating and the money transfers made by young migrants to their families are minimal, thus the commune is losing its reservoir of life force.Certainly the small isolated douars risk a strong regression; and narrowing of the fraction of age of less than 6 years this is why the local and agricultural labour increasingly rare.

3-2 Constraints linked to the conditions and characteristics of the physical environment

Due to its mountainous nature, the site on which the commune of Sidi Boulkhalf is built presents a certain number of constraints which must imperatively be taken into consideration in any operation or development plan for the commune in question. These constraints include:

<u>*3-2-1 an uneven topography influencing the future extension of fagglomerationg*</u>

The topography of the Sidi Boulkhalf commune is uneven and can therefore constitute a serious constraint and an unavoidable factor to be taken into account in the development and future expansion of the urban area.

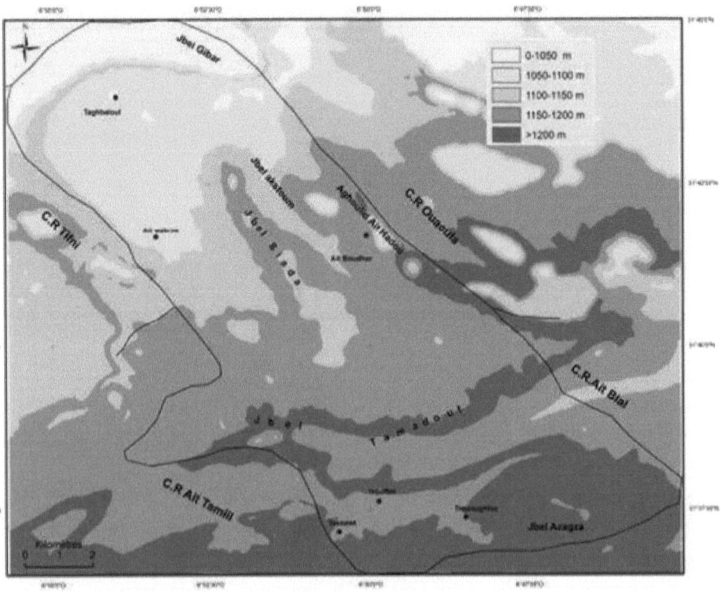

Figure 7: relief map of the commune of sidi boulkhalf showing that the area is mountainous par excellence.

Photo 5: At douar Isquifen (A) and douar Irohan (B), steep slopes and risk ***of landslides***.

Despite the presence of significant forestry resources and the existence of significant potential surface water resources (oued Ghazefe), in addition to the presence of springs, the commune of Sidi Boulkhalf still suffers from several constraints that hamper its development. A mountainous terrain that is unfavourable to agriculture and the predominance of poor soils (60% SAU Hamri, 20%SAU Biad and 20%SAU Harch) plus the problem of annual frost for a 4-month period (December-March) cause serious damage to agricultural yields.

Snowfall over a period of 4 months (December-March) isolates the mountain douars at high altitude during this period. In addition, the legal status of building land through the succession and inheritance procedure leads to a lack of control over the land.

3-2-2 WATERCOURSESANDCHAABA :

Watercourses and chaaba can be a source of danger, particularly during periods of heavy flooding, as they can carry a lot of water and other materials that can cause flooding, the scale and scope of which are difficult to assess, especially in these times of climate change. This is why particular attention needs to be paid to the development of areas close to watercourses and chaaba, in order to prevent the risk of flooding to which they may be exposed.

3-3 constraints on basic facilities and infrastructure

Basic infrastructure is the foundation of all local development in rural areas. They facilitate the establishment of social facilities, housing, trade and industry. The lack or inadequacy of these facilities in rural areas slows down the development process and encourages the rural exodus to other, better-equipped centres and towns. The commune of Sidi Boulkhalf suffers from a glaring lack of basic infrastructure facilities, which is a real obstacle to its development.

3-3-1 Sanitation:

The efforts made as part of the programme to connect the entire population of the municipality to a drinking water supply in 2012 by ONEP from the Hassan I dam, and the mobilisation of associations to manage drinking water, do not meet all the desired ambitions, In the absence of a communal liquid sewage network, the majority of the commune's population uses septic tanks or cesspools as a means of disposal, or else the waste is disposed of in the open air. This has serious consequences for the population's quality of life and the environment (groundwater pollution). These include the absence of a development plan for the commune, plus the lack of a sewage system, which leads to groundwater pollution.

The commune of Sidi Boulkhalf needs a liquid and solid drainage system. However, the commune's financial resources are too weak to help fund this major project. As far as solid drainage is concerned, it should be pointed out that household waste is collected by the population itself, as the centre of Sidi Boulkhalf has no solid drainage system. As a result, household waste is often dumped on wasteland (as shown in the photos).

Photo 6: Household waste is disposed of in the open air, which has an impact on air pollution and the water table.

3-3-2: schooling

No one can deny that the development of every country and every nation depends on the quality of its education, i.e. an effective and diversified education that meets the expectations of planners and decision-makers as well as the needs of the schoolchildren who will be the managers of tomorrow. This objective can only be achieved if we have an equation in which the left-hand side is the pupil and the teacher, and the right-hand side is the platform needed to ensure that this relationship continues. In rural areas, the education situation can only be described as precarious, given the considerable backwardness of the system, despite the efforts made at all levels.

a - Preschool education: There are no preschool education experiments in the commune, but the commune's teachers are fully aware of the importance of preschool education in the continuation of primary education, and see the need to make associations and parents aware of the use of this education.

b- Koranic school: The Koranic school has been the main ëtaЫissement for several years, helping to tackle illiteracy, in particular the school which was set up in douar Tighaduine, authorised by the Habous ministry in 2008, with a total of 60 pupils from neighbouring communes, catered for by the douar's inhabitants. Also the presence of another ë^к coranique having the local to the marabout Sidi Boulkhalf, the total of the bënëficiaires is 40 ëlëves. There is also a shortage of accommodation for the number of pupils at the Tighaduine douar Koranic school.

c. Primary education

The commune has 3 school groups and 13 school annexes, with all the douars within the 2 km coverage zone. Children from the mountainous douars have difficulty accessing school facilities, mainly because of the lack of improved tracks, wadi flooding4 in some douars and wadi flooding in others5.

Almost all children have easy access to the education service (1 km), except in the case of children from mountainous douars, due to the lack of improved tracks and the flooding of wadis and ravines during the rainy season, which has an impact on the children's learning quality.

In terms of infrastructure, 13/16 schools have accommodation for teachers, most of which are micro-satellites. The lack of fencing in some central schools and some school annexes and of caretakers is one of the main factors in the poor state of school facilities and infrastructure, as in the case of some school infrastructures.[7 8 9]

Teacher absenteeism or tardiness is particularly a problem in the mountainous douars due to the lack of access to these douars during the winter period (December-March), the inadequacy or poor condition of teachers' accommodation, the absence of improved tracks in these douars and the lack of transport from the urban centre of Demnate to the douars. As a result, work in the fields during the harvest period, as well as the presence of snow between December and March, especially in the douars of the Ait Toutline fraction, results in pupil absenteeism and a lack of extra-curricular work.

The failure to adapt school curricula to the level of pupils and the rural context, and the

[7] Ghobdade, Irouhane, Ilagaute, Tafraout and Amalou,

[8]Aghri, Tizi and Tagnite.

[9] Ikiss central school, Ikhachene annex, Tirika annex,

irregularity of visits by inspectors, which has a negative impact on the quality of the education service. Not to mention the absence of a specific annual budget for the development and maintenance of school infrastructure and equipment. As a result, the quality of the education service is average or even poor, in particular because of overcrowding or multi-grade classes, teacher absenteeism and poorly equipped school infrastructure.

These problems also lead to children dropping out of school and parents choosing not to send their daughters to school.

d. Secondary education: The commune has no secondary school. The distance from all the douars to the Demnate secondary school (12 to 50 km) is an obstacle to schooling, particularly for girls. In addition, despite the presence of Dar Taliba and the boarding school in Demnate, most girls from the douars in the commune do not benefit from it, given the limited capacity of this Dar Taliba, which is intended for all girls from the rural areas of the Demnate Cercle.

Although the low income of the parents to ensure the expenses related to the schooling and the absence of school transport and daily use of clandestine transport by the pupils to go to the college by paying on average 20 Dhs/day, except the case for the pupils resulting from the douars of the Fraction Ait Toutline whose majority does not continue their studies with the college because of the defective state of the tracks in addition to the remoteness of these douars leads to the increase in rate descolarisation,In addition, the fact that the grant has not been made generally available to pupils who have moved on from primary school to secondary school, the limited capacity of Dar Taliba in Demnate and the lack of school transport between the commune and Demnate have all had the effect of reducing the number of children continuing their studies at secondary school, particularly girls.

To date, there are no literacy classes in the municipality. The literacy programme run for adults has ceased due to the lack of attractiveness of these programmes for citizens; as well as the low allowances offered to trainers and the lack of trained staff to run these programmes.

As far as the dynamic of associations in the education sector is concerned, there are only 3 parents' associations7[10] which are non-functional. In particular, these associations carry out improvement work in schools, with little involvement of parents in central schools and ancillary schools, due to the lack of awareness-raising and capacity-building for these associations to make them aware of their role in improving the quality of education.

The absence of parents' associations has a negative impact on the quality of education, which is why more needs to be done in terms of communication to generate more interest in the development of the school.

3-3-3 Health :

In the commune as a whole, health facilities are limited to a communal health centre in douar Taghbaloute (with one doctor and two nurses). And a dispensary in douar Irohane (with just one nurse).

[10] Ikiss Central School Parents' Association, Louaridene Central School Parents' Association and Azlague Central School Parents' Association.

The health centre in the Taghbaloute douar is staffed by a general practitioner assisted by two nurses, for the inhabitants of the centre and the douars. For more demanding care, people have to go to Demnate, particularly for childbirth, so the dispensary is not supported by a pharmacy and does not have the necessary logistics.
The table below illustrates the precariousness of the health infrastructure and the inequality between the health centres and the demographic map.

***Table 7*: Health coverage based on Delegation provinciale de sante data, 2012**

Distance between douars and the health unit	**Percentage of the population concerned**
Less than 3 Km	14,02%
Between 3 and 6 Km	11,35%
6 to 10 Km	15,65%
+10 Km	58,88%

This disparity in health coverage led us to focus on the main shortcomings, such as the fact that only 68/44 douars fall within the 3 km coverage zone. The population of certain douars9[11] in the neighbouring commune of Ait Tamlil come to the Ait Toutline dispensary for treatment, as does the population of the douars in the Karouele fraction of the commune of Ait Majden, who come to the Louaridene health centre for treatment. Most of the douars are difficult to reach, due to remoteness and lack of transport, as well as roads that are not passable, and wadi flooding during the rainy season in the mountainous douars.
The lack of human resources (only 1 doctor and 2 nurses) is one of the major handicaps to improving health services, as a result of which women from mountainous douars have to give birth in their homes because of the lack of transport and the absence of a birthing centre at the Sidi Boulkhalf health centre, using aromatic and medicinal plants as treatment. And in the event of complications, women are delivered at the hospital in Demnate, Kelaa Sraghna or Azilal. This has led to an increase in the number of women dying during childbirth at home or on the way to the Demnate hospital, especially in the douars of the Ait Ziad fraction, due to the poor state of the tracks and the remoteness of the area.
In conclusion, the existing health infrastructure is far from meeting the demographic weight of the municipality and there is a lack of necessary health equipment (paramedics, ambulance, etc.).
There are 2 to 3 traditional midwives per douar, who often work in critical sanitary conditions that can threaten the health of mothers and newborn babies.
It should also be noted that some people use traditional medicine.

3-3-4: Road network :

The road network in the Sidi Boulkhalf commune is still very modest and insignificant, which has a negative impact on the distribution of products and goods and the movement

[8] Imzizoine, Ait Sigmine, Tialaouine, Irotane, Tangakhte and Timdugha.

[11] Ait Saaid, Igourdane, Imarkassene and Imazayan.

of people, particularly during floods and the winter season.

The diagnosis showed that the road network remains poorly developed, with all the tracks having no improvements and poor to very poor traffic conditions, and almost 30% of them being impassable during the winter months, leaving the populations and the areas they serve completely cut off.

This situation has resulted in a remarkable shortfall in terms of access between douars, with no access to all the main routes from the main town to other douars.

***Table 8*: Condition of the road network in the commune of Sidi Boulkhalf, based on data from the Provincial Delegation of Equipment and Transport**

Road classification	**Registration**	**Link**	**Type**	**Road conditions**	**linear**
Roads Rëgionales	RR302	Imminifri- commune Ait Bllal	asphalted road	good	20 km
Local roads	9 tracks	linking Oued Ghzefe to the Ait Toutline fraction (Ihoujine, Azlague, Tizi, Imaghnasse and Tilsekhte)	prepared track	bad, especially during the rainy season	30 Km
		links between RR302 and Douars Ikhachene, Oukta, Aghri, Ait Yahya, Ait Ouakrim, Amzaourou and Ilaguate	prepared track	bad, especially during the rainy season	12 Km
		links between douar Amalou and douar Hdafa	prepared track	bad, especially during the rainy season	5 Km
		links between douar Ifrougane and douar Azlague	prepared track	bad, especially during the rainy season	5 Km
		links between Tilsekhte and Imaghnasse douars	prepared track	bad	5 Km
		links between douars Ait Boudhar and ghobdade	prepared track	bad	4 Km
		links between douars Ikiss and Ait Aadi	prepared track	bad	4 Km
		linksentredouars Isquifane, Taghadjorte, Tabahannate, Timzughine	prepared track	bad	4 Km
		links between Imzizoine and Igouzoulne douars	prepared track	bad	2 Km

(Azilal)

According to this table, the road network is poorly developed in this commune: only one provincial road crosses the commune; only certain douars (3/44) are close to the road (Taghbaloute, Tighaduine and Ait Sigmine); the other douars are accessible by tracks which are often in poor condition, especially during the rainy season.

Photo 7: View to the east. Regional road 302, which links the Sidi boulkhalf commune and the Ait Blal commune, is gouged throughout due to the extreme vulnerability of the terrain to run-off and falling boulders.

3-3-5 Lack of socio-cultural and sports facilities

In terms of development, socio-community facilities play a fundamental role in stabilising the local population and improving their standard of living, particularly those who are vulnerable or have modest economic conditions. In this way, education, training, health and socio-cultural needs must be constantly integrated into any development strategy.

Infrastructure in this area is clearly non-existent, with the Sidi Boulkhalf has no social, cultural, sporting or service facilities.

Conclusion:

The municipality of Sidi Boulkhalf has significant deficiencies in terms of basic infrastructure, including drinking water, electrification, sanitation, health and the road network.

The relative lack of basic facilities (education, health, sports) is more qualitative than quantitative (lack of staff and equipment, lack of staff housing, etc.).

The centre needs to make up its shortfall in basic facilities in order to be able to meet the new needs that its future expansion will generate. In other words, it is a question of taking into account both the current needs of the population and those required by the future development of the conurbation. Combining these two levels of need is essential if the locality is to increase its attractiveness and competitiveness and improve its position within its immediate environment.

In short, basic facilities still fall short of the average required to raise the municipality to a level with all the qualities needed to provide a decent and dignified living environment. In other words, the shortfall from which this municipality suffers calls for more consistent action to give greater impetus to the dynamism it is displaying. This action needs to bring together citizens and elected representatives in a participatory approach aimed at involving citizens in strategic decisions while applying the principles of governance.

III chapter

A regional territory, focusing on enhancing the

and cultural heritage

Introduction :

The aim of this chapter is to make a concrete project proposal for the commune of sidi boulkhalf. Based on the diagnoses established earlier, the aim is to formulate one or more proposals for an ecotourism and revitalisation project for this commune, based on its shortcomings and shortcomings, but also on its many assets. The project formulated for the commune is based on the human, cultural and landscape potential of the study area, but also on its decline and lack of infrastructure. The project also sets out the support measures that need to be taken, based on experience to date and on reference projects.

It goes without saying that a project that responds to the expectations of the people is much more likely to be accepted than one that only satisfies the rationality of the promoters.

A distinction is made between projects of general interest, which require participation for a limited period of time.

In fact, if the call for participation persists, there will be disaffection and reluctance, particularly with regard to human investment; also, projects where the partners can derive individual benefits are more participative than those where the benefits must be shared; then there are projects that are provoked or aroused, which are more interesting for participation than those that are imposed. This is because induced projects, once they have taken root, leave enough initiative for the people involved; and finally, projects that produce short-term effects compared with those that only produce them in the medium or long term, as the latter require a major material and financial support system.

The expected aim of this chapter is to propose a territorial project that is appropriate to the particular characteristics of the commune and that will aim to improve the incomes and living conditions of farmers in the project area through actions to protect natural resources combined with an increase in the productivity and profitability of farms and more rational use of pastoral resources. It also aims to improve the living conditions of rural women and young people in the area, by promoting income-generating agricultural and non-agricultural activities and basic infrastructure.

The territorial project that we are going to propose will to a large extent involve all the players in the development process, including the local population, who form the backbone of such a development project.

The dynamic nature of the project will enable the Sidi Boulkhalf rural commune to embark on a new process based on a participatory approach, as well as the adoption of strategic planning aimed at carrying out projects, giving priority to the following areas:

1- Forging the basic infrastructure, a solution for an arsenal of problem

According to the interviews we conducted, "tracks are a priority in the requests of the population and the economic and social interest of opening up the area is obvious" (all interviewees). The population is planning to build tracks first, in particular some urgent tracks such as the one linking several douars. This track exists in reality, but it needs to be tarmacked as it passes through most of the douars (almost 22 douars). This track, which is 26 km long, will open up the area for the population in question and will also help to improve civil, health and commercial logistics. But these tracks must not just be tarmacked, they must also be well maintained to cope with climatic and topographical constraints.

The lack of experience of some of the contractors to whom the municipality has subcontracted projects is aggravating the situation and adding to the municipality's budget bill. Hence the need to delegate the task to specialists in the field. The population's contribution is important. It has taken the form of the transfer of strips of land and participation in the work.

1-Water supply is an essential priority for the residents, despite the population's lack of interest in this programme. The local authority must take responsibility for raising public awareness of the urgent need to cover all homes with drinking water, and the need to build water treatment centres, which will help the population in the long term to prevent pollution and certain diseases. This project can be carried out by appealing to associations whose objective is to finance projects aimed at bringing the rural world out of its marginalisation. In addition to this, there is the ongoing mobilisation of support from local political and institutional players (local and communal authorities, local government departments, rural development NGOs, etc.), as a necessary condition for the implementation of the project....), as a necessary condition to ensure the sustainability of the management formula to be adopted, the dëveloppement of local management through the involvement of the population within the framework of community management. This project requires the development of home connections in the douars. This action is in response to the population's growing demand for individual connections.

While the most urgent basic service is access to drinking water, some douars would like to improve their service by opting for home supply.

ONEP remains open to this type of service, provided that certain technical, financial and management conditions are met.

2- Preserving natural resources to reduce poverty

***A - Soil conservation and erosion control*:** Man, through inappropriate practices on the slopes, is the main factor influencing the intensity of erosion. The clearing of forests and natural rangelands, overgrazing, the careless cultivation of erosion-prone land on slopes, mechanised ploughing on steep slopes and the failure to return to the soil the nutrients removed by cultivation facilitate run-off and hence erosion and its undesirable effects on the environment and the economy.

Soil erosion results from the interaction between static and dynamic factors. The static factors are related to the vulnerability of the land. This is a characteristic specific to the environment, dependent on the nature of the land and independent of dynamic factors. The dynamic factors are the pressure agents, which can be either natural (climate and vegetation cover) or human.

Precipitation triggers the process of water erosion, while vegetation limits this process, which means that climate has a destructive effect and vegetation a protective effect. Depending on the way they carry out their activities, human beings can have a positive or negative impact on the process of water erosion.

With the increase in population, the mechanisation of agricultural work, the extension of cultivated land and the degradation of vegetation and soil have progressed alarmingly. The degradation of the vegetation cover has resulted in increased water erosion.

In general, in the Ghazef valley, on slopes that are too steep (due to erosion or overgrazing), runoff is rapid and intense, and is organised in a variety of ways depending on the slope, the nature of the materials and the surface roughness, and attacks the soil at increasing depth, carrying the sëdiments to the outcrop of the bedrock. The process continues until a gënëralisation of the gully and the formation of badlands.

The project must include stone alignments and low walls combined with the maintenance of fertility through organic manure: low walls are small walls built of dry stone (without cement or plaster) along contour lines. They are used to clear plots of land of stones that hinder their development, reduce run-off and its speed, and trap transported sediments. On medium to steep slopes, progressive terraces are quickly created as a result of water and mechanical erosion. As a result, the soil improves in terms of humidity and depth, and is therefore more productive. They are better suited to steep slopes (>15%).

The project must provide for the installation of ridges, which are small strips of earth: following the contour lines, they give the soil a roughness that facilitates infiltration and slows run-off. This type of ridge system allows a maximum amount of water to infiltrate the soil, making it possible to grow a wide range of species. Species requiring a large amount of water are grown in the troughs (alfalfa, for example), and species that are more resistant to drought are sown on the ridges (mathews, for example).

The municipality has also embarked on an initiative to install gabions on gullied areas of concentrated erosion. However, the objectives of the work and its scope are poorly understood, local people do not seem to be involved, feasibility studies are behind schedule and negotiations with local people are dragging on over the choice of work sites. It should be noted that the anti-root treatment will only achieve a high degree of effectiveness if the mechanical treatment is supplemented by a biological treatment over large areas upstream.

B - Rehabilitation of the forest-pastoral estate: this project aims to improve forest-pastoral land by sowing, fertilizing and planting shrubs. It will affect sources of life that are essential to the local populations, who have been subjected to abusive, tolerated or illegal harvesting of leaf fodder, firewood and timber. However, improvements cannot be made without a defence plan, with or without the agreement of the local people. This project must therefore provide for the preparation of concerted forest plans with users with a view to participatory management based on "delimitation of the forest estate, regularisation of illegal occupations, reforestation, improvement of rangelands and socio-economic infrastructure". Water and forestry engineers and technicians must combine their efforts with the local population and authorities to preserve the region's natural resources and deal with the poachers who destroy the forests.

C - Rehabilitation of small irrigated areas : The objectives of our project are the rational development and management of land and water resources, through the agricultural development of large and small irrigated areas.

An appropriate action plan with the following objectives:

- Increase the performance of irrigation systems within the area and guarantee the sustainability of the equipment;
- Rationalise water use.
- Improving the quality of the water distribution service for irrigators;

This action plan focuses on three areas:

- Improving the performance of irrigation infrastructure by stepping up maintenance and rehabilitation of irrigation works and equipment;
- Optimising the application of water in the field and introducing new, more water-efficient irrigation techniques;
- Strengthen the supervision and organisation of users and make them aware of the need for better, more efficient irrigation management.

D - Technical support for farmers: Training is an essential tool for improving the productivity and competitiveness of the target sectors. Its main objective is to improve the technical level and performance of the various operators along the entire value chain, particularly the marketing link.

For growers, the training introduces good practice in plantation management, harvesting and storage. For processors, the aim of the training is to improve processing and value-adding techniques and to introduce a quality management system.

It should be pointed out that the expected objectives will only be fully achieved and the expected results will only be sustainable if the training strongly integrates the human resources of the management structures and professional organisations which already exist or which will be created during the implementation of the project.

The project must therefore provide for the implementation of a training programme that meets the needs of the various categories of stakeholders involved in the target sectors, namely :

- Farmers and their children
- The women.
- The managers of cooperatives and producers' associations, by training project leaders and providing them with technical support, will benefit from their experience in setting up projects, planning, monitoring, evaluation, etc. This will enable project leaders to have a long-term vision to avoid failure.

II- Promote and extend mountain tourism in order to saving the diverse heritage of the sidi Boulkhalf commune

Heritage is a group of assets recognised as such by the local authority in question. The latter gives it a value, linked to its past, which it wishes to pass on to its descendants. These are assets, tangible or intangible, one of whose essential characteristics is to enable a link to be established between generations, both past and future. Heritage is therefore linked to a legacy to be passed on, stemming from the more or less ancient history of the area under study. In this sense, heritage necessarily has a collective dimension. Its conservation is therefore in the public interest. It is a collective good in the economic sense of the term. So, like any good, heritage has a value as a resource that can contribute to the development of the area that produced it (Choay 2009).

The presence of a significant heritage is likely to attract new residents and new activities and therefore contribute to a favourable demographic trend. Admittedly, measuring this attraction effect is tricky. But it can be learned by surveying new arrivals about their reasons for moving to the area.

The higher the level of education and training of the population, the more one of the

essential conditions for territorial development is met. The number of years of education and the highest level of qualifications achieved are the usual indicators of this situation.
If local heritage is valued in local teaching practices, it is a positive factor in cultural enrichment. This can be measured by the number of extracurricular activities based around this heritage, conferences and cultural events triggered by its presence.
A region's level of income depends on its ability to retain the income generated by its own production, as well as that generated abroad, which it is able to attract. Most of the indicators that can be used in this area at a territorial level come from tax data on household income. In fact, regional accounting studies, like national accounting studies, are relatively few in number and almost non-existent at finer territorial levels, as they are fairly cumbersome to carry out. The contribution of heritage in this field comes from its economic value in all its forms.
Firstly, there is the income generated by tourism (Cuvelier P. 1994). They are traditionally measured by income from hőtelière activity, catering, the souvenir trade, visits etc. The presence of secondary residents, attracted to the area because of the existence of this heritage, increases the income of building craftsmen, local businesses and local tax resources.
The volume of employment erëë by the existence of a heritage is an indicator e!ë of ëtudes of the impact of the latter on the territorial economy (ARP-PACA 2008). These are direct jobs such as those liës a l'accueil des touristes (catering, hotels, guiding, transport, heritage maintenance, souvenir shops...). This data is relatively easy to collect. But it is also a question of the jobs induced by the expenditure made in the area, the holders of these direct jobs and, more broadly, all those who benefit from the income created by the presence of this heritage.

1- Protection and enhancement of geological sites with dinosaur footprints:

Unlike biological species, geological ëlëments do not reproduce and the dëtërioration of a site results in its permanent loss: conservation and protection of certain sites are therefore essential to understanding and conserving the surrounding biodiversity. Indeed, gëosphëre and biosphere, geosystems and ecosystems are linked. Today's ecosystems are just the last frame in a film that the geologist is trying to recreate. The geological environment and the Earth's history provide clues to understanding the evolution of life and biodiversity today.
In May 2000, the Association for the Protection of Morocco's Geological Heritage (APPGM) was founded by a large number of Moroccan and foreign geologists. Following several study trips to the province of Azilal, the members of the APPGM confirmed the importance of the regional geological heritage and the need to integrate it into the process of economic and social development throughout the Tadla-Azilal region.
As a result, there is an urgent need to protect and restore these dinosaur sites. The municipality must take certain measures, namely

> Erecting fences so that the sites can be approached without being trampled,

> The construction of shelters against climatic hazards, and in particular insolation and run-off. They are to be built in keeping with local architecture and must blend into the landscape.

> The slabs bearing the footprints of dinosaurs must be cleaned by removing the rubble and any surface debris,

> Cracks, fissures and diaclases must be cleaned by removing all plant roots and rock debris,

> Cracks and fissures are filled with cement for large cracks, and with pumps and syringes injecting resins based on alumina microfine, azurine or glass powder for medium to small cracks,

> At the end, the prints are coated with a mixture of charcoal, lime and glue to make them more resistant and distinctive, and therefore more valuable (APPGM).

> The construction of a wall parallel to the direction of the tracks which will "screen" the prevailing winds to minimise the erosion of the tracks caused by the wind transport of the sands. On this wall, serving as a support, is planned the construction of a "solar ëcraп" by a "parabolic" roof system; the whole of this shelter must be exëcutë in local matëriaux.

Figure 8: Fingerprint protection model. aims to reduce the degradation of its footprints and boost the commune's budget (APPGM).

1- Threshing floors, an integral part of the Moroccan Atlas village landscape

The threshing floors are a major feature of the anti-atlasic terraces. In any case, it is through these structures, and perhaps even more so through those that can be found almost everywhere, here and there, close to the remains of settlements and essentially the old collective granaries, that we can, in a way, bear witness to the agricultural past (Ыё and especially barley) of this Berber mountain. The simple fact that all the Berber dësignes the Ыё by the same term irden* is in itself an indication of the ancientetë culture of this cërëalale in North Africa (S. Chaker, 1995).

In most cases, each family has its own threshing floor. They are built horizontally, as far as possible, close to the douar; they are sometimes even attached to the houses, thus limiting or facilitating the transport of grain as much as possible. However, as soon as the space available for building is Плн1ё on the hilltops, the threshing floors are built on the slopes just below the dwellings and at mid-slope, on slopes sometimes exceeding 60% inclination. They thus share space with the terraced terroirs; both are then subject to the same racesste of adaptation to the slope and both are built with the same lithic matëriel. The stone slabs for the threshing floors are covered with a simple clay mortar to ensure a perfect ëtanchëitë and protect the anrar from the тёГайз of rainwater.

Where space allows, threshing floors are set up side by side, close to the dwellings, taking advantage of the rare slopes offered by this mountain. But as soon as space is at a premium, the fellahs are forced to build their inraren on the sometimes steep slopes. According to the farmers, these structures are built where they are well exposed to the wind, which allows them to cool down while working, and above all so that the dust and fëtus fly away on one side when they could be placed on the other.

The dimensions of the inraren are also varied. The height of the walls, and more prëcisëment of the outer edge of the construction, ranges between 30 or 40cm and 1 m. As with any object built on a slope, the height is also very closely related to the inclination of the slope; the steeper the slope, the greater the need to raise the wall. In short, the walls of the inraren undergo the same treatment as those of the cultivated terraces.

Photo 8: Irraren, whatever their method of construction (paving or rammed earth), whatever their size, have a hole in the centre called tit nounrar oŭ where a stake called tigjdit in Berber is planted, around which the animals, especially the donkey, swing. It should also be noted that threshing floors without this central stake also exist, but they are rare. Like all

As well as being a built amenity, anrar is also maintained during the agricultural cycle. In the case of collective farmyards, maintenance is the responsibility of the entity that built them.

2- promoting tourism on terraced land, an initiative for local development

The terraced landscapes offer a privileged field for the development of cultural tourism. All the other well-known forms of rural tourism or expressions used in analyses and in political language - mountain tourism, hiking tourism, ecotourism, agritourism, ethnotourism, etc. - could be associated with cultural tourism. There are so many expressions that it is sometimes difficult to draw the line between each of these forms or to define them. Terraced landscapes can be maintained or developed for tourism in a variety of ways. Firstly, of course, by showing visitors the very basis of this mountain agriculture, i.e. the cultivated terrace with all its associated features, i.e. the dry stone wall and the irrigation network (ouggoug, charij, seguia). Secondly, this enhancement will also involve presenting what specialists today call the product or "intangible heritage" (J-M.

Decroly et al, 2006), which brings together, in the proper sense of the term, everything to do with custom, local tradition, ways of life, festivals, folklore, crafts, language, gastronomy, customs and rituals, etc. If we take this second aspect into account, the terraced landscapes can offer an original product and a very rich tourist resource that could be promoted and recognised as intangible heritage or "non-built" heritage, to use the expression coined by J-M. Dewailly and E. Flament. Flament.

Enhancing these landscapes by showcasing their ancestral traditions would be more appealing to tourists. It is likely to attract far more crowds than the presentation of the landscape alone. It would even have the advantage of extending the length of stays.

These resources can therefore be discovered in a variety of ways, for example during day trips or special holidays. We have identified below the elements most likely to create a relevant and coherent tourist offer in the terraced landscapes and thus ensure their preservation for future generations.

A -Developing income-generating activities

Hiking is another activity that should be taken into account. Here we can draw on the experience, or rather the feedback, from the many programmes launched almost two decades ago in the Moroccan Middle and High Atlas. This activity will enable tourists to discover the high valleys on foot, with their unique terrain, gorges and deep-rooted Berber culture. Perhaps the best way to help tourists understand how these mountain people live is to take them to see for themselves how a nouba is performed, and why not, if the fellah(s)' work schedule allows, to attend other agricultural activities (ploughing, harvesting, stoning, etc.).

The accommodation should be in the homes of local people, which will help them to understand the way of life of this peasant population, and then improve their incomes thanks to the accommodation on offer. The sale of handicrafts (carpets, mats) and perhaps the hire of mules could round off the activity. As far as home-stay accommodation is concerned, the problem of welcoming strangers into a house remains a crucial issue for the Berbers, some would say. To remedy this, the inhabitant must set aside a space, a bedroom or two, inside the house. He could also build a dwelling outside, close to his house, avoiding, of course, building a concrete or brick house. In other words, you need to retain and/or imitate the features of a traditional house (stone, earth, straw, etc.) while introducing the necessary comforts to ensure hygiene and safety. The option of staying with local people will only be possible and, above all, successful if this last condition is met. Given the population's attitude towards visitors, we could also imagine developing accommodation in places built close to the douar and managed collectively by the jmaa.

b- guided tour of a terrace construction

Show how to build a cultivation terrace or an anrar, another object often associated with terraced land. For example, each time a tourist passes through, a stone or two is added. Various experiments along these lines have already been tried and succeeded in different countries around the world; in France, for example, dry stone wall reconstruction days are organised, attended by enthusiastic visitors. These courses enable both the restoration of supporting walls and an exchange between trainees from all over the world.

c- moussems on tourist circuits

Make the dates of the tours or scours coincide with the moussem(s), agricultural festivals and souk(s) that are well known in the High Atlas mountains. These different events, which could obviously be attributed to cultural tourism, could be an additional tourist attraction. In my opinion, they represent the most visible showcase for the ancestral traditions of the populations visited. On the one hand, they will enable visitors to immerse themselves for a few hours in an environment different from their own, and to interact with the local population for a few moments during a festive occasion. The weekly souk(s) sell a whole range of products: fresh produce (vegetables, fruit), but above all processed products (olive oil, honey), in a word products gathered from the terraces. In some souk(s), handicrafts are also available.

d- promoting food tree cultivation

The cultivation terraces can also be maintained by promoting the traditional food tree crops on which this mountain agriculture is based. These are often very high quality products, such as almond, olive, carob, vine, apple and walnut trees.

A visit to the cooperatives can be organised to showcase the traditional methods and techniques used to obtain olive oil (harvesting by hand, pulping, sorting and finally extracting the oil). It is also the source of a heritage object, the traditional lmaassert mill. However, conservation measures are urgently needed to save and enhance these artefacts, which often fall into disrepair and ruin, to be replaced by modern olive presses.

Tourists themselves, why not, could take part in the whole transformation process; the aim would be to engage in a dialogue with the tourist in terms of mutual exchanges. A direct sales system could be set up: members would set up an organisation that would enable them to follow the whole process, from harvesting to marketing the finished product.

The same concept can be applied to the almond tree, which also undergoes a series of operations, both traditional and ancient.

Ultimately, thanks to the development of these local products, we will indirectly enhance the very support on which they are harvested namely the cultivation terraces. In a way, this will ensure a certain guarantee of upkeep of these terroirs and consequently their pёrепкё, in the same ташёге way that it will help to revive 1 local craft and amëliorate the incomes of the inhabitants.

As we can see, there are many ways of enhancing the value of the terraced landscapes. Although they are not exhaustive, the aim of all these proposals is to help visitors discover the habits and customs of the people they visit. At the same time, as a secondary and indirect effect, they may help to keep alive these practices and this tangible and intangible heritage. Through these resources, we can see that the High Atlas mountains are not lacking in intërët in terms of tourism en дёгагн! and in terms of the dëveloppement of tourism with a cultural and heritage cara^re de iacon particuSre.

We can then envisage that by exploiting all the assets that bënë this mountain benefits from, namely: (a) the existence of landscape and heritage resources, (b) the existence of agricultural and craft know-how, and finally (c) the existence of a major road network, it would be possible to set up tourist circuits offering rewarding excursions in one, two or even more days' travel, if the whole of the central High Atlas is taken into account.

e- A hostel in the Taghbaloute douar to boost mountain tourism

This hostel will boost the region's accommodation capacity. It will certainly boost mountain tourism in a commune renowned for its natural riches. The hostel will be designed in harmony with the picturesque landscape. The structure will offer comfort and tranquillity to mountain lovers. Guitounes" will be set up in the middle of the olive trees for those who prefer to spend the night outside (Amal TAZI, 2003).

This rural tourism project provides financial support to stimulate rural development and help revitalise the local economy. The aim is to enhance opportunities for people living in rural areas, to discourage them from migrating.

The strategy for promoting rural tourism is to ëquip villages with roads to hostels, lmlectricS, drinking water and sewage systems. Naturally, this will not only benefit tourists, but will also help to improve the social conditions of the inhabitants of the douars in which these hostels are located; it will also reinforce the State's policy of combating rural immigration.

Conclusion

Rural infrastructure is essential for reducing poverty and promoting trade and the marketing of local produce. They are also a means of preventing the emergence of conflicts. Our proposal is along these lines, while at the same time defending the idea that roads, health, education, sanitation, etc. are indisputable priorities.

Since the income of the majority of the population is essentially based on agriculture, our project aims, according to the programmes we have proposed, to improve the productivity and profitability of farms with a view to sustainability so that they can play a greater part in supplying urban markets. This is also the key to growth in rural areas.

It is important to point out that successful maintenance or enhancement can only be achieved if certain criteria are met, essentially linked to good heritage management and protection; it is advisable to avoid clumsiness in heritage maintenance. In other words, we must reject any urban inspiration when refurbishing heritage objects. Such a blunder would surely disqualify the character and heritage dimension of the proposed landscape. They should therefore be refurbished while retaining, as far as possible, the old traditional decor and/or landscape. With regard to the services offered and products sold to tourists, it would be essential to encourage craftsmen and traders to adopt a reasonable and affordable pricing policy. Lastly, to achieve the most profitable economic returns possible, these development operations must form part of concerted programmes between all the local tourism partners and the fellah(s). In other words, it would be a good idea to involve the local populations concerned in this development policy. Emigrants, with their financial resources, could play a useful role in restoration projects.

In a nutshell, we can only talk about sustainable tourism and, more generally, a sustainable development strategy in these high-Atlasic foothills if we are genuinely capable of maintaining this population in the mountains in the short and long term, and guaranteeing an additional income for the fellahs, who would at the same time retain their main activity, farming. This is the most important condition that must be met, because the development of tourism could well jeopardise this traditional agriculture even further. Tourism, with the substantial income it can generate, often with less effort than that obtained on the terraces, could in fact divert farmers from farming. The example described by E. Des Bordes in the Cinque Terre (Italy) is exemplary and instructive in this respect. According to the author, the winegrowers, having discovered the importance of the profits from tourism compared with those from agriculture, turned their backs on the latter to concentrate on tourism, which they considered to be profitable in terms of both money and, above all, quality of life. He writes: "The winegrowers, overwhelmed by the mirage of tourism, forgot to visit their land to tend the vines and raise the crumbling dry stone walls", whereas "visitors came first and foremost to admire these fabulous landscapes, forged over centuries by the toil of men, and to taste the wine that was the result of so much effort and sacrifice" E. Desbordes, (1999).

Furthermore, it should be remembered that these areas cannot be managed without the effective participation of their inhabitants and under the responsibility of certain local operators. To ensure that local people are better involved in the various projects, tourism development in the region must never be the sole concern of exogenous investors or civil servants. The local population must be genuinely involved in drawing up and managing projects through elected representatives, local investors or associations (Tribak et al 2006).

General conclusion

In recent years, Morocco has seen significant political advances, in particular with the introduction of new provisions in the communal charter, a process initiated and concretised by the King upon his accession to the throne. This development gave rise to a new concept of authority. In the same context, the adoption of a territorial approach to development was favoured, involving consideration of all the interactions and political sensitivities specific to the territory. This approach looks at the territory as a whole, taking account of its human density, its social dimensions and its historical sedimentations. Today, municipal planning is based on this approach, as well as on the participatory approach, where it is up to the municipal institution, as the bearer of its Municipal Development Plan (MDP), to involve all the local players in an "organised" way. The participation sought goes beyond information or simple consultation, aiming for effective involvement, both material and moral, motivated by the common interest and needs. This form of participation is likely to ensure the continuity of projects after the programmes have been completed.

The success of the process also depends on the gender issue being taken into consideration, where men and women are actively involved in defining their needs and interests according to their own respective realities. At the same time, strengthening the institutional position of the commune and purifying local political work are two essential conditions for ensuring credible decentralisation.

It is essential to provide the population and local players with the means to take charge of their own development, to participate in identifying their needs, and to contribute collectively to finding solutions and implementing the strategic directions they have defined. This is the guiding idea behind the local development plan. The participatory approach is becoming essential, particularly when local development plans are based on participatory territorial diagnosis and participatory strategic planning.

Consequently, it is clear that all the foundations and principles of municipal planning require effective communication within the plan's governing bodies and, ideally, between the various bodies involved.

Participation must be seen as a means to an end, not an end in itself. Implementing it requires time and effort to raise people's awareness, understand their motivations and work out solutions with them. Any attempt at agricultural and rural development must therefore be a long-term process.

Furthermore, it is becoming necessary to move from an approach that favours local people's participation in projects to one that aims to support local initiatives. To this end, it is vital to ensure that grassroots organisations are supported over the long term, so as to encourage their capacity to integrate into the decentralisation and professionalisation processes. This requires the development worker to be in tune with the local dynamic and to work with the population concerned to successfully anchor it in its regional, national and even international context.

The municipality is undoubtedly the governing body for local development, and therefore has a responsibility to take charge of the communication strategy for establishing partnership and mobilising the other players. It is up to trained and prepared local elected representatives to perpetuate this new culture of partnership for development by promoting closer links between local players, the population and external services in a permanent effort to serve the common good.

For the younger generation, mountain farming no longer holds any significant interest. In their eyes, it's a thing of the past. Unlike previous generations, who did not necessarily view work and effort in terms of productivity, today's young people place the emphasis on profitability. Their unanimous discourse refutes a reality: a considerable investment in terms of work throughout the year generates only very modest returns. They question the value, and above all the profit, of this activity. Their perspective is not completely wrong, because a few months spent in the city offer other employment opportunities that are much more remunerative than a year's work on the farm.

Most agricultural work is still carried out in the traditional way, using human muscle power: fields are ploughed with a plough and/or hoe, and cereals are harvested with a sickle. Even older people describe this approach to farming as outdated, describing it as "old-fashioned".

Faced with this situation, it becomes imperative to look for alternatives and implement solutions to revive fellahs' interest in their landscapes. Although a number of possible uses can be envisaged, the one that seems most promising for reviving these landscapes is the development of a heritage tourism activity. In this context, the terraced landscapes could constitute a key support for cultural tourism in the central Moroccan High Atlas.

However, for this initiative to be successful, it is essential that the terraced fields are henceforth recognised by the public authorities and the local population, particularly the farmers, as a cultural heritage to be preserved and enhanced. In addition, it would be beneficial to combine this enhancement with other historical, landscape, architectural, social and economic elements specific to the High Atlas mountains.

In the medium and long term, this approach could improve profitability for farmers. If successful, it could even help to attract new populations to this mountainous region, or at least retain those who currently live there.

All in all, if tourism is to be beneficial and sustainable, it needs to be taken in hand, for the benefit of the local populations, and once again this is only possible if the local populations themselves take charge of it, in close collaboration. This is true with the local communities and authorities, the services of the Ministry of Tourism and of course with mountain tourism professionals, and even with all the associations and other civil society organisations that have the mountains as their area of action.

References

* ♦* Amal TAZI , Taza : Une auberge pour animer le tourisme de montagne PибHë dans L'Economiste le 05 - 09 - 2003 , Article disponible en ligne a l'adresse suivante http://www.maghress.com/fr/leconomiste/50167 (page consul le 23/06/2013)

* ♦* ARP-PACA (2008), L'impact ëconomique et social du patrimoine en region Provence- Alpes-Cote d'Azur, Cahiers de l'Agence regionale du patrimoine.

* ♦* AUBERTIN (C.), 2001, "La montagne, un produit du dëveloppement durable", Revue de Gëographie Alpine, Grenoble, n°2,

* ♦* Choay F. (1992), L'altegorie du patrimoine, Seuil Paris.

* Choay F. (2009), Le patrimoine en question, Seuil, Paris.

* ♦* Cuvelier P. (1994), Torres E and Gadrey J., Patrimoine, modëles de tourisme et dëveloppement local, L'harmattan, Paris.

* ♦* DESBORDES Emmanuel, 1999. Les paysages de terrasses agricoles dans le bassin mëditerranëen septentrional: ëtude comparee des Alpes-Maritimes et la Ligurie.

* ♦* IFAD (2001), Participatory approach for an impact-oriented project cycle - workshop report "Strengthening the impact orientation of IFAD's project cycle" Rome.

* ♦♦ Jacques Gandini (2000), Le Geoparc du M'Goun, Article available online at http://www.prehistoire-du-maroc.com/iroutlane-iwariden/Page-3.html (page consulted on 10/06/2013)

* ♦* J-M. Decroly, A-M.Duquesne, R.Delbare and A.Diekmann, 2006. Tourisme et sociëtë: mutations, enjeux et dëfis, Bruxelles, Editions de l'Universite de Bruxelles, p. 66.

* ♦* J-M. Dewailly and E. Flament, 2000. Le tourisme, Paris, Collection Campus, Ed. S.E.D.E.S, p. 98.

* ♦* M. Ait Hamza. 2000. "Environnement et stratëgies paysannes dans le Haut-Atlas central", In : A. Laouina ^d.), la montagne marocaine, dynamiques agraires et dëveloppement durable , Rabat, Universite Mohamed V, Publications de la chaire UNESCO-GAZ NATURAL " Gestion de l'environnement et dëveloppement durable ", p. 48.

* ♦* Monographie de la commune sidi boulkhalf : CT 18-03 de Demnate, May 2003

* ♦* NAJIM A et VEDELAGO F (2001), l'agent de dëveloppement local, Emergence et consolidation d'un profil professionnel, Chaire Unesco Universite de bordeaux 3, Ed La lause, Bordeaux.

* ♦* RGA 1996, veterinary service DPA Azilal

* ♦* S.Chaker, 1995. "Linguistique et Prehistoire : autour de quelques noms d'animaux domestiques en berbere", In : L'homme mëditerranëen, Mëlanges offerts a Garbiel Camps, R. Chenorkian (dir.), Aix-en-Provence, Publications de l'Universite de Provence, p. 259.These de doctorat en Geographie, Universite de Limoges, 2 volumes, 430 p.

* ♦♦ TRIBAK Abdellatif, LOPEZ LARA Enrique, M. Jose Miranda BONILLA, LAAOUANE Mohamed (2006) ,Tourist activities and sustainable development in an area of

MIX
Papier aus verantwortungsvollen Quellen
Paper from responsible sources
FSC® C105338

Printed by Books on Demand GmbH, Norderstedt / Germany